BEI GRIN MACHT SICH IHR WISSEN BEZAHLT

- Wir veröffentlichen Ihre Hausarbeit,
  Bachelor- und Masterarbeit

- Ihr eigenes eBook und Buch -
  weltweit in allen wichtigen Shops

- Verdienen Sie an jedem Verkauf

Jetzt bei www.GRIN.com hochladen
und kostenlos publizieren

Mukasa Aziz Hawards

Fundamentals in Nutrition Health

GRIN Verlag

Bibliografische Information der Deutschen Nationalbibliothek:

Die Deutsche Bibliothek verzeichnet diese Publikation in der Deutschen National-
bibliografie; detaillierte bibliografische Daten sind im Internet über http://dnb.d-
nb.de/ abrufbar.

Impressum:

Copyright © 2007 GRIN Verlag GmbH
Druck und Bindung: Books on Demand GmbH, Norderstedt Germany
ISBN: 978-3-656-50137-4

Dieses Buch bei GRIN:

http://www.grin.com/de/e-book/233341/fundamentals-in-nutrition-health

Name: Mukasa Aziz

Course Name:

Fundamentals of Nutrition Health

Table of Contents

Course Description

The course in environmental impact assessment concentrates on the scientific aspects of food, emphasizing reasons for procedures used and phenomena occurring in food preparation.

The course also elaborates the public health orientations in the regulation and policy that pertains food supply, quality assessment and assurance.

Objectives

- o To understand the scientific foundation of food science
- o To recognize food pairing and that food choices in meal management have nutritional, cultural, psychological, economic, and safety influences
- o To understand sensory and objective food evaluation
- o To comprehend the scientific principles utilized in the preparation of carbohydrate, protein, and fat rich foods as well as in the complex system of batters and dough.
- o To obtain proficient understanding of the regulations, policies and laws that concerns food supply.

Literature Review

Nutrition and health are coherent when it comes to maintaining or supporting the living human systems (lindlahr 1940). The health related complications in humans if we are to consider proximate estimations arising from different clinical analyses show a shocking percentage of about 80% of these arising from matters to do with nutrition.

Though our study in nutrition and health is not mainly based on this shocking trend it is an international aim to pursue health and if the methods undertaken are to be realistic then there is a crucial need to extend our observations and strategies towards the nutrition factors that are characteristic to all human societies. The further shocking issue in the global nutrition trend is the serious attention towards nutrition in the old aged, infants and nursing mothers than in the youth and the ages between 25 to 50 years.

As the study will show from the various world surveys, that these are not the only nutrition issues causing tension but there still exist more far reaching factors that grievously shift the basic nutrition patterns of humans to the non-healthy patterns. Among these factors are; levels of education, economic status, cultural and ethnic influence, religious influence, the food industry plus policy and regulations as controlled by the regional governments. Perhaps one of the chief factor or goal that was supposedly to be addressed in the millennial world campaign by the world states in some of their conventions had to be the nutrition though as our we know few or completely no convention was aimed at addressing this problem but instead politics and world economics have always dominated these conventions.

The science of nutrition shows that our body systems are dependant of the daily nutrients ingested and these substances act as the building blocks for our living systems, energy producing factors and great contributors towards body maintenance and repair. The study of the fundamentals in human nutrition and health therefore, opens us to wide range of topics that address most of the key issues in the nutrition detailing the body metabolism and the socio-economic factors in this criterion.

Chapter One
Basic Concepts in Nutrition: Nutrition, Kilocalories, and Nutrients

Nutrition

Nutrition science is a new subject introduced to man's attention after the evolution of biology and chemistry. However, it is established that nutrition gained its greatest momentum within the 20th century during which the scientists outlined tow equivocal phases in characterizing this science.

In the starting phase which also acted as the foundation underlying the research in nutrition science, they were able to discover and characterize the various vitamins in which case their deficiency syndromes were described more in details. Also during this initial phase the recommended and required dietary nutrients were analyzed after which they were estimated according recommended dietary allowances (RDAs).

Additionally, in 1997 the recommended dietary allowances were further formulated using volumes containing dietary reference intakes which included extreme quantities of nutrient intake beyond the recommended ones for a health persons to such amounts that tough excessive but they present no detrimental effects to health in man. This also on the contrary characterized the second phase of the idealized modern day science of nutrition (Goldman: Cecil medicine 23rd edition).

The second phase conceptually concentrates on the dietary reference intakes (DRIs). These in the modern perspective of nutrition science develops evidence that ramifies the dietary relationships alongside the nutrition profile as main precursors for diseases that have stigmatized the health complications within the international community for instance cancer, and the coronary heart disease of which these two are haunted as the principle cause of silent death chiefly in the western states.

The term nutrition refers to the science that involves the study of nutrients in food their composition, role and metabolism in the living system. Therefore, nutrition is not enclosed to food substances and diet but it also broadens in the analysis of the body physiology as the various food substances channels through the digestive tract after ingestion and its final destination during absorption and assimilation.

Nutrition science seems to be taking a basic position in health sciences as it turns up to resolve most of the questions about a certain number of body complications. In this case, the health professionals have progressively been able to conceptualize certain ideas and theories as they tend to be interconnected with nutrition issues while solving the critical threats posing challenge against health wellness in man. The evolution of modern science have experienced the emergence of theories and concepts that from time to time has strengthened the principle of nutrition in science study and these have also been further supported by experimental analysis demographic survey results, and more so the real life experiences as seen in different individuals.

The scientific discoveries that have proved the fundamentality of nutrition in human science have no basis on superstitions as it has been the case in science such as the big-bang theories but rather more visible and non-fictions (Thompson, 1910).

This is so because nutrition derives its support and chief position in human science on tangible evidence, which is not limited to laboratory experimentations but even much more in the common environment around us. Our lives are key addresses to the modes of our nutrition and to be more frank it can be asserted that our health status act as mirrors through which we view or observe the discipline in our feeding. These days it has been so much materialized for instance in the case of obesity that the outward appearance in weight, skin texture and other observable features as symbols of nutrition orientations in individuals which scientifically may not be the right case yet when analyzed this concept and assertion may 60% be confirmed true clinically. However, before jumping to non-scientific conclusions there is a gap that we should consider vital and be left filled which actually opens us to some fundamentals in the process of feeding in man; its pillars, pathway and the role it plays in human existence.

It should be noted that feeding is not restricted to animals but instead to all the five kingdoms of the living things, as they have been outlined with plants and bacteria inclusive.

A brief overview of feeding in organisms

Feeding according to nutrition refers to the intake of food substances from the surrounding environment that contains a nutritious value into the body of an organism. Remember that in this case the term organism is being used in its primary denotation where it means a living thing hence asserting that feeding is a life process general for all the living organisms.

From scientific points of view, feeding has been categorised into two forms; autotrophic and heterotrophic which sometimes are also used to distinguish between plants and the animals though this pattern of classification have of recent been dropped due to the fact that though some organisms are not plants they show distinctive features that are not possessed by animals either. The autotrophic form of nutrition refers to the ability of an organism to synthesize its own food by using the available natural resources and this is rendered to the green plants. The green plants through the process of photosynthesis are able to build up food substances from common natural elements such as carbon, and elements of water by the use of solar energy absorbed by the green

pigment; chlorophyll which is located in the chloroplasts of their cells. The light energy from the sun is used to build up bonds between the constituents elements that combine to form a food compound that can either be simple or complex.

However, unlike the autotrophic form of nutrition in plants, the animals exhibit a more dependant form of nutrition i.e. heterotrophic nutrition in which case the organism lucks the biological mechanisms of synthesizing its own food. The animal in this case therefore, can only use other organisms for its sources of food and nourishment due to its inability to convert the available natural elements into food compounds.

The main difference that offsets the position of these two distinct categories of organisms in feeding originates from the cytological differences.

The protoplasm of the plant cell is mainly saturated with the chloroplasts which act as cellular pockets for the chlorophyll pigment that plays a vital role in the absorption of light energy emitted from the sun whereas that of the animal cell is completely colourless due to the absence of chloroplasts. Some non-plant organisms such as some bacteria however, contain a particular type of chlorophyll that enables them to trap light for synthesis of organic compounds in the process called chemosynthesis. These bacteria also derive energy from chemical processes around them, which they then use in the formation and manufacture of food substances.

The pathway of food through the gastrointestinal tract

The animals take in food substances by ingestion through the mouth from where the substances are subjected to breakdown both chemically and mechanically in the process generally referred to as digestion. The process of digestion can either be physical, in which case the complex ingested food substrates are only broken into small molecules that can further pass through the slander channels of the alimentary canal (also known as the digestive tract). In this case, there is no altering the chemical nature of the ingested food. This physical digestion involves processes like;

- Mastication; which is the chewing action of the teeth as supported by the voluntary upward and downward movements of the jaw bones to grind and crush food on the surfaces of the teeth.
- Peristalsis; which are the wavelike contractions and relaxations of the esophagus and the intestines to soften food and propel it in the forward direction for further processes.

Food from the mouth after being acted on by the chewing action of the teeth is the sent to the esophagus due to the pressure exertion by the tongue and the further support by the roof of the buccal cavity, which also forms food into a ball-like substance, called a bolus. During swallowing, the larynx that opens into the trachea or wind pipe is protected by the glottis which is covered by the movements of the epiglottis so that no solid particles enters the respiratory tract. All this protective processes for the delicate tissues of the respiratory tract is involuntary under the control of the cerebrum of the brain.

While in the esophagus food is pushed forward ways by the wave-like motions of the walls and opened into the stomach by the cardiac sphincter muscle. The stomach walls secret the gastric juice that contains;

- Hydrochloric acid; produced by the goblet cells and it provides the acidic medium for the protein enzymes and disinfects food from all germs and disease causing bacteria.
- Mucus; produced by the oxyntic cells and it lubricates the walls of the stomach during contractions and movements. Mucus also protects the walls of the stomach from self-digestion by the protein enzymes.
- Protein enzymes; produced by the peptic cells and they carry out the chemical digestion of the protein food arriving in the stomach by converting it into peptides, and casein.

However, chemical digestion also takes place within the mouth for starch food as effected by the salivary amylase enzyme breaking and converting starch into maltose sugar. After physical and chemical digestion in the stomach, the pyrollic sphincter then releases food shortly towards the duodenum for further digestion of the undigested food. The pancreas organ release the pancreatic juice after stimulation and the juice contains the following substances;

- Pancreatic amylase; which is an enzyme that digests carbohydrates and continues the digestion of maltose sugar which had been previously terminated by the acidity of the stomach.
- Pancreatic lipase; which is intestinal another enzyme that breaks down the lipids into fatty acid molecules and simple glycerol.

The enzymes are chemical substances made up of organic constituents mainly proteins in nature and their role is to catalyze or speed up the rate of metabolic reactions in the body with digestion inclusive. The enzymes are produced from goblet cells of the gut and their activity is specific in nature and mediums of PH.

The different food substances are digested in specific regions of the gut in which a suitable PH medium can be located for the enzyme activity such as the main digestion for proteins that takes place in the stomach due to the acidic medium that is favorable for protein enzymes.

Absorption and assimilation of food in the gut

Both the physical and the chemical digestion of food renders the molecules a small size and an absorbable state that can pass through the small internal pores till the reach their final destination for use or storage in the body.

The process of absorption takes place in small minute molecules of the various food types such as the monosaccharide glucose for carbohydrates, amino acids for proteins, and fatty acids/glycerol for proteins and these have different channels of passage during their absorption though the ileum or small intestines play the greatest role in the process of absorption. Hence the main importance of the enzyme is realized at this stage as determinants for the particular and specific food molecule to be build up from a particular food type and therefore their absence in this criteria will present a considerably vast inhibition in the general body metabolic progress.

However, it should be noted that not all food types are digested in the body but some are directly absorbed across the walls of the gut without digestion. For example the mineral ions, vitamins and water whereas the dietary fibers or roughages are not digested in the body chemically due to their chemical composition that require the cellulose enzyme for digestion. The cellulose-digesting enzyme is not found in human guts hence the roughages are egested in the same chemical nature though their physical nature is always altered by physical processes caused in the gut.

The ileum is adopted to the absorption function due to the following anatomical features;

(a) It is a very long slander tube structure with the length approximating to about 2meters long a feature that provides a large surface area for the absorption of the already digested food molecules in the endothelium.
(b) The ileum is also internally made up of finger-like projections called the villi, which mainly allow the absorption of the end products for lipid and carbohydrates. The villi are suitably made of thin epithelial covering layers that allows easy passage of food molecules with very little resistance. The villi are further surrounded and covered with the microvilli, which also propel food during absorption to make the process quick and fast.
(c) In addition to the above is the juicy medium located inside the lumen of the ileum that allows the dissolving of the semi-solid food particles there by converting them into the flexible liquid state that can easily be absorbed via the pores of the lumen walls. The juicy medium in the ileum is contributed by the intestinal juice or succus-entericcus also secreted by the gut walls and it contain the digestive enzymes which finalize the digestion of all the ingested food substances.
(d) Finally yet importantly is the high supply of blood vessels around the walls of the ileum, which allow the close circulation of blood in which the absorbed food nutrients are transported and taken to their final regions of use and storage.

In assimilation, the already absorbed food substrates are the incorporated into the body metabolism to their specific roles in the body. However, the process of assimilation is a selective one involving the final instruction implementation of the body cells to decide the food type to be

used immediately, stored and deaminated as it is the case for the excess amino acids and the quantity of molecules to be stored in this case.

For healthy living, it is recommended in nutrition procedures for the body to be subjected to all the various food values though these should be provided in controlled amounts in order to prevent body complications arising from obesity. Scientific research has provided evidence showing that the amounts of food substances required for the different food types are non-equivalent but vary from one type to another though analysis goes further to show that these quantities are estimates consumed at the basal metabolic rate (BMR). This means therefore that relative quantities of food molecules used or require by the body may still decrease or increase in accordance to factors of activity, temperature, body demands and the health status of the individual.

Kilocalories

Most of the different food types particularly proteins, lipids, and carbohydrates release energy on their breakdown in the body systems while the vitamins, mineral ions, water and roughages have minute or completely lack energy units in the body (Winton, 1945).

The energy released by on breaking the food molecules is measured in small units called kilocalories or Calorie (C). A kilocalorie is a measure of energy and this is commonly heat energy emitted on burning a weighed quantity of the food substance. A kilocalorie like a kilogram that contains 1000grams also contains 1000calories which are normally denoted with a "c" of the lower case though the media and most of the packing material for foods and beverages this has been form of denotation has been misused.

However, this negligence though not given much attention it can be a result of serious implications most especially for individuals who are under strict diet recommendations as medical procedures for reviving their health wellness. Take for instance in the case of infants and the plagued diabetic victims where specific amounts of food calories are recommended per day. In this case the use of packed and tinned food for preparing meals will present a stout challenge and at the same time a risky one because there is no provisional method that can immediately be undertaken to determine whether the packing company meant a Calorie "C" or a calorie "c". in this course study however, this rendering has been treated with the recommended denotations where a kilocalorie is the same as the Calorie denoted with "C" in which case this contains 1000 calories "c".

Each energy releasing food type has a particular amount of energy units released in an individual at rest and awake. The energy needs of the body are measured using the basal metallic rate (BMR) which is the energy consumption of a living person at rest. During the resting period most of the body systems are still active to support life though rate of their functionality is relatively low as compared to that for the person carrying out vigorous activity such as athletics.

The ongoing processes during rest include heartbeat to support circulation substances in blood, movements of the thoracic cavity to support gaseous exchange, control activity of the brain to support other processes and so on.

The estimated amount of kilocalories an individual requires is dependant of three main factors namely;

1. The energy requirements of the living body at rest; this is the basal metabolic rate and it refers to the amount of energy consumption to support the processes active in a resting individual.
2. The degree of physical activity of the body; the amount of Calories required in a physically active body is higher than that required in the resting state.
3. Thermic effect of food; this refers to the amount of energy required by the body to digest and absorb food for assimilation. This sometimes is referred to as **Specific Dynamic Action** (SDA) and it is equal to approximately 10% of the total daily Calorie intake (Winton, 1932).

Estimating the Basal Metabolic Rate of an Individual

The basal metabolic rate of an individual (resting metabolic rate) is mainly influenced by his or her genetic composition, height, weight, age, skin surface area, and the level of activity. For example female person aged about 20years with at least 160 cm tall (5'3'') weighing about 50kg (110pounds) has 1.5 square meters of skin-surface area. The energy production in Calories (C) released per square meter of skin varies with a person's age and sex.

We can assume in this criteria that this woman on a daily basis uses about 886C for each square meter of skin and therefore her basal metabolic rate is estimated to be 1,329 kcalories per day i.e.

$$1.5 \text{ X } 886 = 1,329 \text{ C.}$$

Where,

-   1.5 is the value of the skin surface area
-   886 is the amount of kilocalories used on a daily basis by the individual

(To convert pounds into kilograms, divide the body weight in pounds by 2.2. and to convert height in inches to centimeters; multiply inches by 2.54)

Generally, the basal metabolic rate also varies from individual to individual due to differences ranging from:

-   Body height
-   Gender
-   Age

- Growth rate and status
- Level of exercise
- Health status of the individual
- Use of drugs such as tobacco and caffeine

From experimental evidence, a remarkable trend is presented showing high levels of basal metabolism in males than in females when other factors are kept constantly normal. In the graphical representation below **(figure.1)**, the different basal metabolism is compared for different two patterns of exercising for a normal human (moderate physical activity and vigorous physical activity). The activities under observation include;

- Dancing
- Walking
- Stretching
- Playing golf
- Weight lifting
- Hiking
- Bicycling
- Yard work/ light gardening

Figure.1: showing the amount of kilocalories consumed in a 154-pound person in the basal metabolic rates for different activities for both moderate and vigorous physical activity

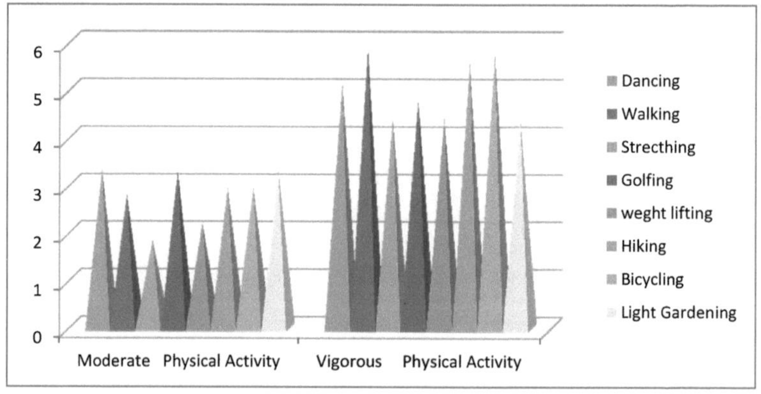

Extracted from the Food and Nutrition Board (US-Government) www.iom.edu/subpage.asp?id6634

Nutrients

This refers to components of food substance that have a nutrition or food value and are capable of nourishing and building the living body.

There are over fifty elementary nutrients of food though all these are mainly grouped according to six categories, which include;

1. Proteins
2. Carbohydrates
3. Lipids
4. Vitamins
5. Water
6. Mineral ions

The nutrients are the actual "food values" that provide importance to the living systems with the main of these being production of energy that support life processes in the body and effect maintenance and repair of the cells.

These food nutrients are required in various quantities by the body and for this reason they have also been categorised into macronutrients and micronutrients. The term "macro" means large and in its completeness it has been used to refer to food nutrients required in large quantities by the body whereas micronutrients is used to refer to food nutrients that are required in small quantities.

Food types such as carbohydrates, proteins, lipids and water are complements of macronutrients while the vitamins and the mineral ions are required in smaller quantities hence forming the group for the micronutrients. However, the macronutrients are not for all the life span of human are they used in the same quantities but these values have a threshold point at which they become detrimental and induce health complications on the body, take for example the sugars or carbohydrates which may result into diabetes or the excessive fats which cause obesity.

Therefore, even the macronutrients have a lifetime at which they are required and after it they are then needed in smaller quantities. The food nutrients have also been categorised into energy yielding and non-energy yielding food substances.

The energy yielding food substances include the carbohydrates, lipids and the proteins while the non-yielding food substances include the vitamins, water, and mineral ions. However, this method of classification has developed a pseudo attitude of less importance of such substances which in actual sense is not true. These non-energy yielding food substances are of great value to the body in various ways which are not necessarily energy related yet without them the body cannot perform and exist properly.

The energy values are measured in Calories per gram of the food (a gram is unit of mass on metric scale though not the standard unit) and the following kilocalories (C) have been actualized for the energy yielding substances;

1. Carbohydrates releases 4 kilocalories per gram
2. Lipids releases 9 kilocalories per gram
3. Protein releases 4 kilocalories per gram (www.healthfinder.gov)

The carbohydrates are the major energy yielding food substances though according to analytical data they release less energy than the carbohydrates. This attribute is caused by the respiratory quotient used in determination of the basal metabolic rates of these food compounds.

In the chemistry perspective these nutrients are rendered a classification mainly depending on the presence and absence of the carbon element. According to chemistry, compounds are regarded as organic if they contain carbon in their composition such that if we are to borrow this rendering then food substances such as carbohydrates, proteins, lipids and vitamins which also contain carbon fall in this category while inorganic substances are those that lack carbon such as the mineral ions and water.

The basic functions of the various food nutrients in the body are also grouped into kinds of energy production, maintenance and repair and regulatory roles just as summarised in the table below (figure.2).

Food Nutrients	Function Of Food Nutrient In The Body		
	Regulatory Functions	Energy Production	Induce Growth And Body Maintenance
Proteins	✓	✓	✓
Carbohydrates	✓	✓	✓
Lipids	✓	✓	✓
Vitamins	✓	X	✓
Mineral ions	✓	X	✓
Water	✓	X	✓

Source: Wiley, Harvey Washington. Not by bread alone; the principles of human nutrition. New York: Hearst's International Library Co., 1915

Carbohydrates

These are the organic food compounds with the main function as energy yielding substances in the body.

In so many cases the term carbohydrates has been mistakenly understood by referring it to table sugar, pasta or potatoes which actually is very misappropriate. This term covers a lot of

compounds though they may all have sugar as the primary component. However, sugar when technically applied it is always referring to mono- or disaccharides.

The monosaccharaides are mainly simple sugars such as fructose, glucose, and galactose while the disaccharides are made up of two monosaccharide molecules and these include;

1. Maltose sugar = glucose molecule + glucose molecule
2. Lactose sugar = glucose molecule + fructose molecule
3. Sucrose sugar = glucose molecule + galactose molecule

However, the category of carbohydrates apart from the simple sugars also contains complex compounds which are called polysaccharides. These contain infinite numbers of simple molecules and they include starch in plants, glycogen which is food storage compound in humans and cellulose that forms the plant fiber tissues.

The basic properties of carbohydrates is their sweetly taste which is mainly for the simple sugars and their ability to dissolve in organic solvents. Examples of common sources of carbohydrates include; milk, maize, rice, wheat, onions, potatoes, fruits, etc.

Starch from plants is the primary form of compound from which we derive the carbohydrates for the body and this or glycogen when broken down in the body they turn into simpler molecules of monosaccharaides (glucose) which provide energy during respiration in form of ATP molecules (Adenosine Tri Phosphate).

Simple sugars are also used in the building blocks for the synthesis hereditary molecules i.e. nucleic acids which make the DNA double helix strands.

 A diet deficient of carbohydrates results in fats being oxidized and used to release the ATP, also excess carbohydrates in the body can be sometimes converted into fats as storage tissue, which can later be broken down to release energy during times of starvation.

Lipids

These are sometimes referred to as fats which I consider to be wrong, a diversion from accurate classification. This is so because not all lipids are fats but rather all fats are lipids in this case therefore, it should be noted that, fats are just one form of the three main subclasses within the lipids i.e. Phospholipids, Steroids and True Fats.

The steroid type of lipids are mainly used in the synthesis of body regulatory chemicals and substances such as the hormones, though important they are not recommended to be added to the normal diet most especially the steroid group of cholesterol that has health implications attached to the causation of high blood pressure disease and heart failure.

The phospholipids on the other hand should completely not be added to the oral diet simply because these are main components in the structure of the living cells where they form part of the enveloping cell membrane.

However, the true fats which are scientifically referred to as **Triglycerides** are of great value to the body and hence they should be added to the normal daily diet. These true fats are capable of releasing maximum energy Calories of about 9 KC which is considerably greater than the value of energy released by both the proteins and the carbohydrates.

Some true fats contain a peculiar type of **Essential Fatty Acids** called <u>Linoleic Acid</u> and <u>Linolenic Acid</u> that cannot be synthesized by the body hence according to the nutrition standard recommendations such food nutrients should always be part of the common meal.

The basic role of the lipids in the body is to release energy and provide body insulation especially during very cold temperature fluctuations. Apart from these two functions, lipids are also components in the structures of the body systems for example the kidney, skin, and other glands of the body.

Proteins

The basic structural units in all proteins are called the amino acids which are linked to each other by the peptide bonds; however, the type of amino acids vary from one type of protein to another.

In nutrition science, proteins can be categorised into two main groups: <u>Incomplete Proteins</u> and <u>Complete Proteins</u>. The complete type of proteins contain all the vital amino acids required for efficient body repair and functioning whereas the incomplete proteins lack certain amino acids. The living human system is capable of synthesizing some amino acids on its own though some amino acids cannot be synthesized in the body this means therefore that they have to obtained orally by eating foods rich in them. The type of amino acids that the body cannot manufacture are called <u>Essential amino acids</u> and these have been outlined below and their common sources;

i.   <u>Threonine</u>; dairy products, nuts, soybean, and turkey.
ii.   <u>Lysine;</u> dairy products, green peas, beef and turkey.
iii.   <u>Methionine</u>; wheat, fish and oatmeal.
iv.   <u>Isoleucine</u>; peanuts, oats, and lima beans.
v.   <u>Tryptophan</u>; sesame seeds, sunflower, poultry, peanuts, lamb, sunflower seeds.
vi.   <u>Arginine</u>; beef, peanuts, ham, shredded wheat, and poultry.
vii.   <u>Phenylalanine;</u> calves' liver and peanuts.
viii.   <u>Histidine;</u> human and cow's milk
ix.   <u>Leucine</u>; fish, soybeans, peanuts and beef

They main function of the proteins in the body is to build body structures and support cell multiplication during growth. This is merely considered as the main function though proteins are used for other purposes such as energy production, regulatory function and body repair. Like the carbohydrates the protein food are capable of releasing 4kilocalories of energy per gram of food ingested by man.

The quantity of proteins required by the body depends so much on the age and the type of amino acids available or needed for efficient body functioning. The actual size of proteins required by the body have been a subject of great contradiction most especially in the diet conscious intellectuals, though to save this discussion it is recommended to have at least <u>50grams</u> of proteins in the body on a daily schedule. This is because the excess amino acids are not stored in the body, therefore, the body has to get rid of almost about 20-30 grams of proteins by the end of the day for proper functioning.

The excess amino acids are never stored inside the body but rather they are eliminated through the process of deamination in the liver after breaking them into urea and water. Proteins act as sources of energy only in the last stages of the body redemption from starvation or during fasting after the metabolism of fats deposits and carbohydrate decomposition.

Mineral Ions

These are elements existing naturally in the environments and cannot be synthesized by our body systems. Minerals are normally obtained as constituents in some food substances and barely can they be ingested in their solid pure forms.

Though the minerals are ingested while dissolved in food solutions and in water they retain their natural characteristics and play roles which differ ranging from being enzyme activators, transmitters, regulators, and controllers of enzyme reactions. Some minerals have a catalytic effect in the chemical reactions of the body and hence adjusting the body's rate of metabolism in which case the operate like the enzymes and in this criteria they are called catalysts because unlike the enzymes which are organic , the minerals are inorganic in nature. The table below contains the list of minerals required by the body, their source and roles in the human systems.

Figure.3: Showing the Various Minerals and Their Roles in the Body

Inorganic Ion (mineral)	Recommended Daily Intakes For Adults	Physiological Value	Readily Available Sources	Other Information
Calcium	1,000mg	Builds and maintains bones and teeth	Dairy products	Infants need 1,300mg; VD needed for absorption
Fluoride	3.1-3.8mg	Maintains bones & teeth; reduces tooth decay	Fluoridated drinking water & seafood	
Iodine	150µg	Hormone thyroxin	Iodized table salt & seafood	Some soils have iodine content so iodized salt is vital
Selenium	55-70µg	Used in enzymatic reactions	Meat, grains and seafood	
Iron	15-10mg	Composition of hemoglobin	Legumes, poultry, dried fruits, meat & grains.	Females need more than men; pregnant women need 2 times the normal dose
Magnesium	320-420mg	Bone mineralization; muscle & nerve function	Dark green vegetables, whole grains & legumes	
Phosphorous	700mg	Acid/base body balances; enzyme cofactor	All foods	Infants need 1,250mg; most people get more than recommended.
Zinc	12-15 mg	Wound healing; fetal development, involved in various enzymatic & hormonal reactions.	Meat, fish & poultry	

Adapted from: Elements of the theory and practice of cookery; a textbook of household science, for use in schools. New York; London: The Macmillan Company; Macmillan & Co. Ltd., 1906

Water

Water is a binary compound needed in all body processes for proper metabolism that support an efficient life in humans. according to scientific analysis and biochemistry it is estimated that almost 90% of all the body composition is mainly made up of water. Therefore, due to tis factor water takes the principle importance in the body than the rest of the other food nutrients.

Water neither yields energy nor builds the body but yet without it, none of all these processes can take place after all it is water that provides the medium in which these substances are hydrolyzed and dissolved for circulation throughout the entire body.

Though excess quantities of water in the body presents no direct adverse effects to the human systems it is still important to regulate these quantities because they alter and offset the physiological equilibrium of the internal environment which either slows the rate of metabolism or inhibits enzyme activity. Some of the properties of water that renders it suitability in the internal environment include;

- It has two dipole moments hence forms both positively and negatively charged particles, a characteristic that enables it to conduct heat and electricity in the body.
- Pure water is neutral in PH. i.e. it is nether acidic nor alkaline hence forms solutions with both PH orientations
- Has a high specific heat capacity hence a good compound to effect temperature regulations of the body.
- Water also has a very high latent heat of vaporization hence it is a good agent of cooling the high body temperatures.
- Water is also considered as a universal solvent and hence capable of dissolving all forms of substances; both organic and inorganic.
- Pure water is colourless hence it adds no components to solutions it forms due to pigments.

❖ **Functions of water in the living systems**
i.   Water is used in the activation of body enzymes which are chemical in nature. Water breaks the bonds in the reactive sites of the enzyme molecule such that the food substrates are able to fit and occupy the sites.
ii.   Water provides the medium for the transportation of substances in the body both excretory wastes for elimination and the nutritious compounds to the regions of need in the body.
iii.   Water is a regulatory substance used in the control of abnormal temperature fluctuations in the body most especially by evaporation while the body from heating up.
iv.   Water maintains the buffer state of the blood tissue such that the internal environment is not offset from normal functioning during metabolism.

Vitamins

These are organic food nutrients required in small or minute quantities in the body metabolism. Vitamins exist in numerous types a reason that actually presents an intellectual challenge in determing their function in the body. However, one feature that vitamins possess in common is that they are consumed or required in very small amounts.

Just like the essential amino acids and the fatty acids, vitamins cannot be manufactured in the body except vitamin D which in in a scientific point of view is not a vitamin. The saga arising from the position and nature of vitamin D is simply solved by acknowledging the scientific facts. The fact is that this vitamin is not really obtained from the sun (ultra violet rays) as it was assumed at first but rather it is scientifically proven that these sunrays only strike the cholesterol type of steroids lipids in the skin causing it to yield and turn to vitamin D.

Though still upsetting to establish the general function carried out by vitamins in the body, it has come to our attention that these minute compounds have a repairing role they play hence asserted that their main and principle function is to promote and boost body immunity against disease.

There have been a considerable negligence in recommending the actual quantities of vitamin that should be taken into the body for proper metabolism regarding that may be for the case of vitamins the issue of quantity is not paramount as it is in the other food types therefore, asserting that a person can as much as he can without stopping. This however, is misconception and can lead to health implications because actually much quantities can be fatal for some types of vitamins. Categorically we can group the vitamins into water soluble and water insoluble type of vitamins.

High dosage of some water soluble vitamins such as vitamin B6 (pyridoxine) have toxic effects to the internal environment of living systems where it causes nervous complications such as numbness of the hands and unsteady gait.

Other functions of vitamins is that some of them act alongside body enzymes in metabolism as coenzymes since they are organic in nature just as the enzymes.

Chapter Two

Interaction of Nutrition with Health and Disease

Impact of Nutrition on Health

In 400.BC Hippocrates, the father of medicine once complemented to his students on the importance of nutrition in a healthy living when he asserted that they should let their food be their medicine and at the same time their medicine to be their food. Objectively, this might not sound coherent with the medical professionals who may view such an advise as a diversion that tends to threaten their service market though this is just my brief perspective that avails less in the current health dynamics (Sutherland, 1925).

There has been tremendous evidence confirming the interrelatedness between health and the diet orientations within individuals from the various researchers and health scholars. These connections are even above the detrimental levels of health threats that the world has never realized before ranging from chronic disease and fatal circumstances. However, these connections are not limited to disease but they open more channels for the influx of other risks including social and behavioral variables and for this reason therefore, there is stout demand for more extensive studies and research to establish where the possible strategies should be implemented.

First and fore most, it should be noted that the connections and interactive trends between health and the nutrition aspect have been extracted from some profound epidemiological surveys but yet these too should not be regarded as final as they may neglect or be unable to consider casual relationships. This sometimes poses questioning and provides room for uncertainty as such variables presented merely confounded due to the fact that they were not first examined to balance the dynamic changes from one unit of survey to another.

These epidemiological case studies also face another disgrace as it is asserted in one perspective that they barely consider the diet assessments in free living individuals. Though this seems to be dubious one principle fact to be perceived is that diet trends orientations certainly are individual patterns and in only limited circumstances can they be group characteristics. Considerations should be undertaken on an individual basis also because most of the factors that drive diet selection in individuals are so much distinctive for instance;

- Flavor
- Education

- Religious and spiritual factors
- Cultural and behavior
- Health status
- Activity and occupation
- Finance and economic status
- Emotional influence .etc.

Scientific studies on animals can be applied to avert some of these limitations caused by epidemiological data but they too are vulnerable to direct confrontation and criticism arising from experimental outcomes that may certainly be different from that encountered by humans. Prospective studies on humans have been carried out in the most recent years on the diet orientations at a random basis to establish the disease risks that have their root and precursors arising from individual diets. Some of these studies have left pending questions and created curious demand for more analysis while others have been positive to the objective.

The nutrition problems that cause devastating effects on the health status of individuals and leading to nutrition related disease prevalence at an international scale can be grouped into <u>Individual Diet Behaviors</u> and <u>Associative Nutrition Factors</u>. These two categories of nutritional disease source factors are responsible to most of all the common disease related to nutrition.

The individual diet behavioral establishes that some health complications are so much personal originating from an individual's eating discipline and life style. As stated previously it is true that the flavor of a particular dish can taste different from one person to another or even equally the same but yet this does not mean that it is healthy for everybody. Individually people are adapted to special meals and foods even when they are of low nutrition value in relation to their body needs and in this case this kind of specialty on a single less advantageous meal a person misplaces the attention on the more valuable dish. This factor also goes beyond diet to lifestyles and routine behaviors that an individual is used to for instance exercising and level of physical activity.

Individuals who give attention to exercising even for nonathletic intentions show a high degree of fitness and tend to reduce health complications most especially those culminating from overweight and excessive body fats. However, it is always advised to exercise while on a recommended diet for both athletic and the nonathletic persons as this makes the outcomes more positive than negative.

Fitness is not actually a characteristic of a healthy life but it only increase the chances to attaining a health body and reduce on vulnerability of the body to disease contraction.

The individual diet behaviour can be an allergic or chronic factor within the individual in which case it is hereditary or inherited directly through conception from the parents or relatives. These

kind of individuals have irreversible traumatized health status where by their health is only offset by the ingestion of allergic food substances causing them severe sickness where in some cases of neglect they may have fatal consequences. On this same pattern however, the nutrition scientists have devised eating schemes and recommendations or references to cater for individual genetic differences that have health implications while exposed to different dishes. Among some of these common individual genetic differences are variations in blood groups, chronic abnormalities such as sickle cell anemia, albinism, hemophiliac etc.

Individual diet behavior factor also include the addiction aspect in which an individual may be addicted to particular drug and other addictive substances that cause health risks. Take for instance smoking and alcoholism, these substances have long-term biotransformation pathways in the human system at a particular levels of concentration and hence they may cause severe abnormalities in the body physiology and more inhibit the circulation and purity of the internal environment. Individuals attached to excessive addition of such substances are vulnerable to easy contraction of disease such as cancer, high blood pressure, liver disease and heart failures.

On the other hand, the associative nutrition factors are group based and cause health risks at a group or community level other than on an individual basis. Such factors range from occupation influence, cultural and social behaviours, religion and spirituality, policy and regulations, the food industry, education levels, climate and environment, plus the economic bases of the communities.

These associative factors have a common trend in that an individual is affected even against his choice but due to the given association he is given to adopt the nutritional patterns which sometimes may not be favorable to the health needs of his body. take for instance the occupation influence in given it considered to factory work in low developed countries where an individual earns about 3000/= (approximately $1 and 10 cents) per day. Now this is so common in Ugandan factories most especially those managed by Indian investors and an individual has to work for 8hours per day to qualify for such wages plus a restrained meal that certainly does not meet the body needs despite of the hard work. Such employees are indirectly to adapt to this kind of meal in order to save the little daily earning that can support their families with basic needs of medical acre, accommodation, food and clothing.

The health status of such an individual is under a great tease due to poor nutrition services offered at work and yet even the home gets affected by the same factor.

Cultures and social behavior aspects of some communities have diet orientations that endanger the health status of persons. The cultural norms and taboos of some tribes in Uganda for instance have diet restrictions to some persons claiming that such foods are abominable and non-sacred to some individuals in which case such persons are deprived of their chances to good nutrients as required by the body health requirements. Therefore, such individuals can develop deficiency diseases and other disease risks due to malnutrition, as their health status is not well nourished with the recommended nutrients.

Also religion is another factor among the associative nutrition factors that channel in the nutrition related health implications. Some religions restrict some dishes from their followers referring to them as unclean or abominations before God and hence any individual that ingest foods is banished thereby discouraging the others to do the same act. Though this concept paper has nothing attached to people's beliefs yet since this has showed up, we should not leave it unaddressed.

Mortality and Morbidity Trends Due To Nutrition Imbalance

Around the international community, the population has so much acclimatized to disease resulting from nutrition disorders though little fatal experiences have not yet been widely registered. This however, has made more misconceptions with in the international community that perhaps nutrition disorders and diseases are temporally and may be incapable of causing long-term health implications on the victim but yet this idea is greatly reversed more so by even practical or real life experiences in low developed countries of Africa, Asia and the south America region. Nutrition diseases have caused devastating effects to the livelihoods of the poor individuals from these countries both psychologically and physically in their life status (winters, 1902).

However, these shocking mortality and morbidity rates have not been limited to the low developed states of the world only but they have ramified also into the better off states in Europe and the USA. In the table (figure.4) below, the most common diseases due to nutrition affecting the majority of individuals in the United States have been enlisted with their respective associated factors. For in as much as there is a possibility of reducing the prevalence of these diseases from individuals it is a more creditable plan analyse these associated factors with the decrease and increase of the disease. This serves to redeem the hope within the victimized community or individuals by illustrating how and which type of personal care and nutritional discipline that doesn't call for trained health personnel to implement but that even the can be carried out by the individuals under the pressure.

Figure.4: Table Showing The Dietary Influences On The Major Causes Of High Mortality And Morbidity Rates In The United States Of America

Mortality Cause	Factors Causing Decreased Risk	Factors Causing The Increased Disease Risk
Heart disease	Ingestion of non-reducing carbohydrates, essential fatty acids, soluble fiber, polyphenols, proteins and antioxidants e.g. vitamins E, C; beta-carotene; selenium. Also folic acid and moderate alcohol	Feeding on saturated fats, cholesterol; possession of excessive calories, Na^+; and abdominal distribution of fats in the body deposits
Accidents		Consumption of alcohol beyond manageable amounts and concentrations
Cancer	Fruit and vegetable enriched diet	Excess calories, fats, alcohol, red meat, tined beef, and excessive distribution of abdominal fats
Suicide		Consumption of alcohol and substances of abuse beyond manageable amounts and concentrations
Diabetes mellitus	Intake of enriched diet	Abnormal levels of body calories, fats, and alcohol and excess abdominal fat deposits.
Cerebral vascular disease	Staying on a diet enriched with K^+, Ca^{2+} and omega-3 fatty acids.	Excessive consumption of Na+ and alcohol as related in hypertension.
Hypertensive renal disease and hypertension	Increased fruit diet enrichment, and Ca^{2+}, K^+, vegetables, Mg^{2+} plus omega-3 fatty acids.	Excessive intake of Na+, alcohol, excessive levels of calories, maximum saturated fats and excess fat deposition in the body abdomen
Morbidity Causes		
Alimentary constipation (diverticular disease)	Feeding on a dietary fiber enriched diet.	
Obesity	Regular exercising and low fat food consumption.	Excessive intake of fats and calories
Neural tube defects	Feeding on dishes enriched with folic acid.	
Osteoporosis	Moderate levels of Ca^{2+}, vitamin D, and vitamin K	High levels intake of vitamin A, proteins and Na^+

Source: Goldwin; Cecil medicine 23rd edition (disease influenced by nutrition) and American dietary guidelines

An Overview on Critical Nutrition Disorders

Eating disorders are mainly associated with the mind conception of an individual pertaining the eating habits and from here these turn into psychological trends and traumatic characteristics posing excessive abnormal intake of food or inadequate induced appetites for special types of food or reduced quantities below the required amounts according to the body needs.

Generally, three eating disorders have been regarded common which also have a relationship to the perceptions and cultural values in which we are oriented. The three common ones are; anorexia nervosa, bulimia and obesity (Strouse, 1997). However, there has been an asserted criteria according to the Diagnostic and Statistical Manual of Mental Disorders; 4th Edition (DSM-IV) from the Psychiatric Association of America. In this criterion, only two eating disorders are distinctively considered prevalent i.e. anorexia nervosa and bulimia while commending that the third category is non-specific but instead it includes all the other disorders related to feeding behaviors and discipline.

1. **Anorexia nervosa**:

This type of eating disorder has its root from the fear to become overweight by an individual and mostly pronounced in adolescents and preadolescent women. It is also regarded as a nutritional deficiency disease characterized by severe or prolonged weight loss as a result of a voluntary severe restriction in food intake (Thomas, 1945).

This eating disorder has two subtypes; binge eating/ purging and restricting. The binge eating or purging subtype of anorexia nervosa is characterized by regular engagement in binge eating or the purging behaviour. The purging behavior involves self-induced vomiting or the misuse of laxatives, diuretics or enemas (precursor drugs for bowel evacuation) (British medical formulary).

Restricting subtype of anorexia nervosa, the victim is not involved in any form of purging behavior or binge eating. Persons with this disorder have a distorted perception of their body sizes and shape while comparing themselves to others (Winchell, 1924). They see themselves as excessively fat when, in fact they are starving to death. Society's preoccupation with body image, particularly among the young people, may contribute to this sort of incidence. Common symptoms of anorexia nervosa include the following;

- Weakened muscles
- Hypotension (low blood pressure)
- Hypothermia (low body temperature)
- Inflammatory bowel disease
- Decreased heart rate
- Slowed reflexes
- Reduction in size of uterus

2. **Bulimia nervosa**:

More likely as to do with anorexia nervosa, bulimia is also associated to psychological problems and depression. It involves periodical binge eating character followed by purging of the body by inducing vomiting using laxatives and enemas or prolonged fasting in order to undo the effects of binge eating. Bulimics are usually of normal body weight or overweight.

The following is a list of bulimic symptoms;

- Lethargy
- Diarrhea
- Migraine headaches
- Damage to teeth and gum
- Kidney malfunction
- Extreme mineral deficiency
- Severe stomach cramps

3. **Obesity**:

This disorder occurs when people consistently take in more food energy than is necessary to meet their daily requirements. In one perspective obesity appears to be a simply solvable problem though health wise this presents more challenges than to merely restrict food intake.

In obesity the quantity of food intake may not necessarily be the cause but it may have its roots from genetic origin and characteristics of an individual in which case it is irreversible though its health implications can be averted and suppressed (peters, 1921). The largest population of individuals suffering from obesity are related to a random phenomenon other than to genetic orientations.

Chapter Three
Food Laboratory and Procession, Food Industry, Policy and Regulations

Food Laboratory and Processing

The food industry acts as epicenters in the health determination of the communities because they are responsible for the supply and distribution of the edibles which are basic needs to human life.

Not surprisingly, the successive generation of human development have experienced a tremendous evolution of industrialization in which also the food industry has gained its momentum. Currently, the food related industries and firms place 3rd among the forerunning industries that generate the highest profits in most of the states and when it comes to the multi-purpose supermarkets, you will find almost 60% of the various commodities on shelves related to food.

However, such a broad emergence of food and beverage companies has mainly been boosted by the increased knowledge in nutrition among the population that causes a demand for quality foods and beverages. Such a demand has presented challenges and at the same time opportunities for food companies to invent and improve on the farm products such that they can meet the demands and at the right time (Rose, 1917). New methods have been embedded in food processing to enable long-lusting preservation, more nutrient value and among all these food treatments it is still another challenge that the more formulas applied to the food is the more the food becomes less effective and perhaps with even more health implications.

Food processing means the process of transforming the raw ingredients of the farm product into consumable food or even food into other different forms. What food processing does is to extract the nutritious constituents of the clean farm harvest of both crops and animal meat, converting them into attractive shelve forms of products that can be marketed.

In most of the food processing practices the primary goal to be achieved is to enable the preservation of the food and its nutritious value for long-life durability and if possible with its quality maintained. Some of the key preservation approaches undertaken in the food processing industry are;

- Salting
- Roasting
- Smoking
- Steaming
- Oven baking.
- Autoclaving
- Sun drying. Etc.

However, most of the above methods in food preservation are of ancient origins and with less technological requirements. These methods for instance roasting or smoking could be used to alter the physical state of food such as for flesh and fish in which case the water content is retrieved to minimize the chances of rotting and micro-organism invasion.

The modern methods however, have seen the trend of food preservation change where canning and food tinning is also becoming prevalent in beverage companies. In so many cases today food is not deprived of its water content and this in most cases is regarded unnecessary after all new technological advances allow application of chemical preservatives that can maintain the food status for quite a prolonged period of time most especially in military maneuvers.

This on the side also alters the chemical nature of food by transforming the chemical components responsible for a particular bond structuring in biochemical stability of the constituents and turn the components into other unknown forms of compounds that could result in expected experiences of both positive and negative health impact onto an individual.

Food Industry, Policy and Regulation

The food industry has been subjected to maximum inspection and supervision by the statutory bodies and quality enforcement government agencies in order to ensure the quality requirements and health needs are met in the processing of food (Prince 1939). Though these activities have mainly been developed in the transforming of food companies in the western and more developed states of the world in the low developed and poor states this is not a fundamental achievement that can be seen taking place.

In the under developed states these food companies are not severely inspected to check the quality of food as determined by the factory conditions along together with the methods applied. The main role of inspection is to determine the extent to which these companies maintain safety and healthy assurance aspects at the work place.

The main aspects that are determined in food company inspection mainly include the following;

(a) Health Aspect;

Inspection aims at achieving the necessary analysis on the food company that can be used to assess and evaluate the health risks and the degree of the health implications that the company has towards the surrounding and to its occupants. The surrounding involves the food consumers to which the products are distributed, and ecological units. The inspection process therefore does not stop on assessing the health and physical status of the company but also checks the health quality of the products by determing their nutrition values and considering whether this substance has no health implications to the consumers.

(b) Hygiene Aspect;

This sounds related to health but yet its principle role is vital simply because food is a basic component for life and its state determines the health status of the consumer. Hygiene inspection is normally directed towards the occupation conditions and physical state of the company. What hygiene inspection does is to assess the hygiene status of the working environment and the labour team itself, determing whether the ideal and recommended protocols are met and that the sanitation profile of the company will not be of grievous impact on the quality and health status of the food product.

(c) Efficacy Aspect;

This inspection level determines the effectiveness of the food product in the life of the consumer. The degree of efficacy inspection varies from one state to another though has nothing to do with the recommended values as stipulated by the regulatory bodies in this criterion. Efficacy determination involves analysis of both the quantity and quantity of the product ad this on the consumer's side protects the rights of the consumer.

Quality determination agencies such as the ISO investigates and evaluates the quality standards of the product to ensure that the consumer is not exploited by the food company by indication of incoherent and inconsistent net weight on the packing's or even that the indicated nutrients are actually available in the food product sold.

Policy and Regulation Overview

The legal society in the European community has been exerting pressure on the food processing companies compelling them to advance their hygiene and safety status. This serves to solve to critical problems at the same time (Wiley, 1929).

First of all, hygiene and safety conditions at the food factory ensures conducive work conditions for the labour team and this averts all the possibilities of antagonizing the quality of food products in procession.

Secondly, hygiene and safety promotes and maintains the good health quality of the food products. Hygiene basically ensures that all material used have been sterilized and free of microbiological infections and this is done alongside other technological methods such as autoclaving and homogenization like in the long lusting milk products.
Through the international community policy and regulations have been asserted on the food and beverage industries pertaining the various activities involved in the food procession approach and these include the following;
(a) Registration of the food and beverage company

Before the establishment is erected it is required to apply for Licence and obtain credibility and statutory support from the government authorities such as the trade ministry and registrar of companies. A considerable period is taken in completing the documentation at this level in order for the authorities to ascertain the legality of the company and its owners.

(b) It is a policy throughout the international community to subject the food processing company to inspection before starting the production process. This inspection ensures that the company meets all the recommended and expected standards that determine quality products. In most countries even in Africa inspection is continuous throughout the existence and operation of the company so that its activities do not deviate from the standard operation procedures.

(c) Packaging and distribution

It is a policy regulation that all food and beverage companies indicate all the necessary information on the food product before it is distributed to the consumers in order to safeguard and raise awareness where needed for consumer protection. The necessary information should include the following;

1. Name of the product
2. Name of the company and its address (telephone, website, fax, physical address & email)
3. Manufacturing date and expiry date for the products
4. Net weight or approximate quantity of the packed product
5. Inspection stamp from the government inspection agency
6.   Composition percentage of all the nutrient constituents present in the product.

Chapter Four

Recommendations, Conclusion and Bibliography

Recommendations

For the best outcomes of the food industry and consumer right protection there is still pending need to revise the processing methods by analyzing their advantages varying from one to another and the best of all to be executed. The procession of food is of great advantage towards man in that it ensures that food can be consumed at the right time since preservation is manageable hence there is no fear of wasting and spoiling.

Food processing companies not only reduce the fears of food spoilage and waste but also ensures that food is consumed with the taste and flavor of choice to the buyers and also saves time and energy that would be lost at home preparation. Actually, the benefits achieved from food procession are broad and they cannot all be addressed in this brief study scope however, on realizing its importance it is of great advantage to outline strategies that can promote it and support its quality.

Factually, it is of great importance to consider the role of the different stakeholders in the nutrition aspect and to analyse or deduce the best approaches for these parties to effectively influence the positive trends in nutrition. These stakeholders are the government, food processing companies and the individual consumers of the food products.

Principally the regional governments of the international community should revise the policies and regulations governing the food processing companies for consumer protection and for the best interests of the food companies. The government should address the revenue and taxation trends on food products from the home industries if the progress and prosperity of health and the food company are to be achieved. These taxes fail the food companies so much in that the companies try to improvise by manipulating the consumer in order to redeem the and compensate for the high revenues deducted from their sales (Murray. J, 1916).

The governments should also implement strategies of modern farming into the rural areas in order to promote high yields for the common peasant farmers who actually are the sole providers for the row materials used by the food processing companies.

The health ministries in the regional governments should implement civic education. This ensures the education of the public about the new trends in nutrition and approaches to minimize the eating disorders and addressing other disease risks culminating from nutrition. Balanced diet maintenance, regular exercising, and feeding on organic food should be advocated within the society as some of the possible self-indulgences for individuals to eliminate and minimize sickness influenced by nutrition.

Conclusion

Nutrition value has been subjected to a great test for over the years and various generation in human existence to assess its credibility in life and its relationship with health and all this has been positive. Consider for instance the time of Daniel in the ancient times of the Babylonian empire in which vegetable dishes were put to a test by Daniel's three friends while restricting themselves from meat delicacies served in the palace (Bible; the book of Daniel). The results of this test and human experiment gave shocking deductions that had not been known to man earlier that a vegetable diet can bring forth good health without health implications as compared to the meat diet.

The argument in this issue is that it is important for the international community to realize the importance of nutrition as a key to health living and given much time and attention the poor health disparities related to nutrition will be averted from our societies. For in as much as life is still precious to humans its sustainability should be paramount and therefore all the possible threats against nutrition should be disarmed by all means either through advanced studies and research or through policy strategies by the government agencies.

Bibliography

1. Arend armitage, clemons drazen, grigs larusso: Cecil medicine; an expert consult title. Online print. Published by Saunders, an imprint of Elsevier Inc. 2007. USA
2. Lindlahr, Victor Hugo and National nutrition society: You are what you eat. New York: National Nutrition Society, 1940.
3. Murray, J. Alan: The economy of food; a popular treatise on nutrition, food and diet. London: Constable & co. ltd., 1916.
4. Price, Weston A: Nutrition and physical degeneration; a comparison of primitive and modern diets and their effects. New York, London: P. B. Hoeber Inc., 1939.
5. Rose, Mary Swartz. A laboratory hand-book for dietetics. New York; London: The Macmillan company; Macmillan & co. ltd., 1917.
6. Rose, Mary Swartz: The foundations of nutrition. New York: The Macmillan company, 1927.
7. Sutherland, George Alexander: A system of diet and dietetics. New York: Physicians and Surgeon's Book Co., 1925.
8. Strouse, Solomon and Maude Alice Perry: Food for the sick; a manual for physician and patient. Philadelphia, London: W. B. Saunders Company, 1997.
9. Thompson, Henry: Food and feeding: with and appendix. London; New York: F. Warne, 1910.
10. Thomas, Gertrude Ida: The dietary of health and disease. Philadelphia: Lea & Febiger, 1945.
11. Winton, Andrew Lincoln and Kate Grace Barber Winton. The analysis of foods. New York, London: Wiley; Chapman & Hall ltd., 1945
12. Winton, Andrew Lincoln and Kate Grace Barber Winton. The structure and composition of foods. New York, London: J. Wiley & sons Inc.; Chapman & Hall limited, 1932.
13. Winters, Joseph Edcil. The food factor as a cause of health and disease during childhood or the adaptation of food to the necessities of the growing organism., 1902.
14. Winchell, Florence E. Food facts for every day, for upper elementary and junior high school boys and girls. Philadelphia: J.B. Lippincott, 1924.
15. Williams, Mary Emma and Katharine Rolston Fisher. Elements of the theory and practice of cookery; a textbook of household science, for use in schools. New York; London: The Macmillan Company; Macmillan & Co. Ltd., 1906.
16. Wiley, Harvey Washington. Not by bread alone; the principles of human nutrition. New York: Hearst's International Library Co., 1915.
17. Wiley, Harvey Washington. The history of a crime against the food law; the amazing story of the national Food and drugs law intended to protect the health of the people, perverted to protect adulteration of foods and drugs. Washington, 1929.

18. Peters, Lulu Hunt. Diet and health, with key to the calories. Chicago: The Reilly and Britton co., 1921.